A Whale's Tale

Wyatt's Antarctic Adventure: Tagged by Scientists

Written by Dr. Reny Blue Tyson
Illustrated by Dr. Gail Tyson

Dedication: To future scientists, explorers and curiosity lovers.

ISBN 978-0692375860

BlueSTEAM.org

This story is based on research conducted by real scientists in Antarctica. The research was supported by the National Science Foundation and the United States Polar Program (grant: ANT-07-39483) and was permitted under the U.S. Marine Mammal Protection Act by the National Marine Fisheries Service (permit: 808-1735), the Antarctic Conservation Act (permit: 2009-014) and the Duke University Institutional Animal Care and Use Committee (permit A041-09-02).
More information about the research this story is based can be found by reading the following scientific publication: Tyson et al. (2012) Synchronous mother and calf foraging behavior in humpback whales *Megaptera novaeangliae* insights from multi-sensor suction cup tags. Marine Ecology Progress Series. 457: 209-220.

Meet Wyatt, a baby humpback whale who loves adventure and making new friends.
Like other baby whales, Wyatt is called a calf.

Wyatt was born six months ago off the coast of South America.
His mother, Wendy, takes great care of him.
She is a wonderful mother.

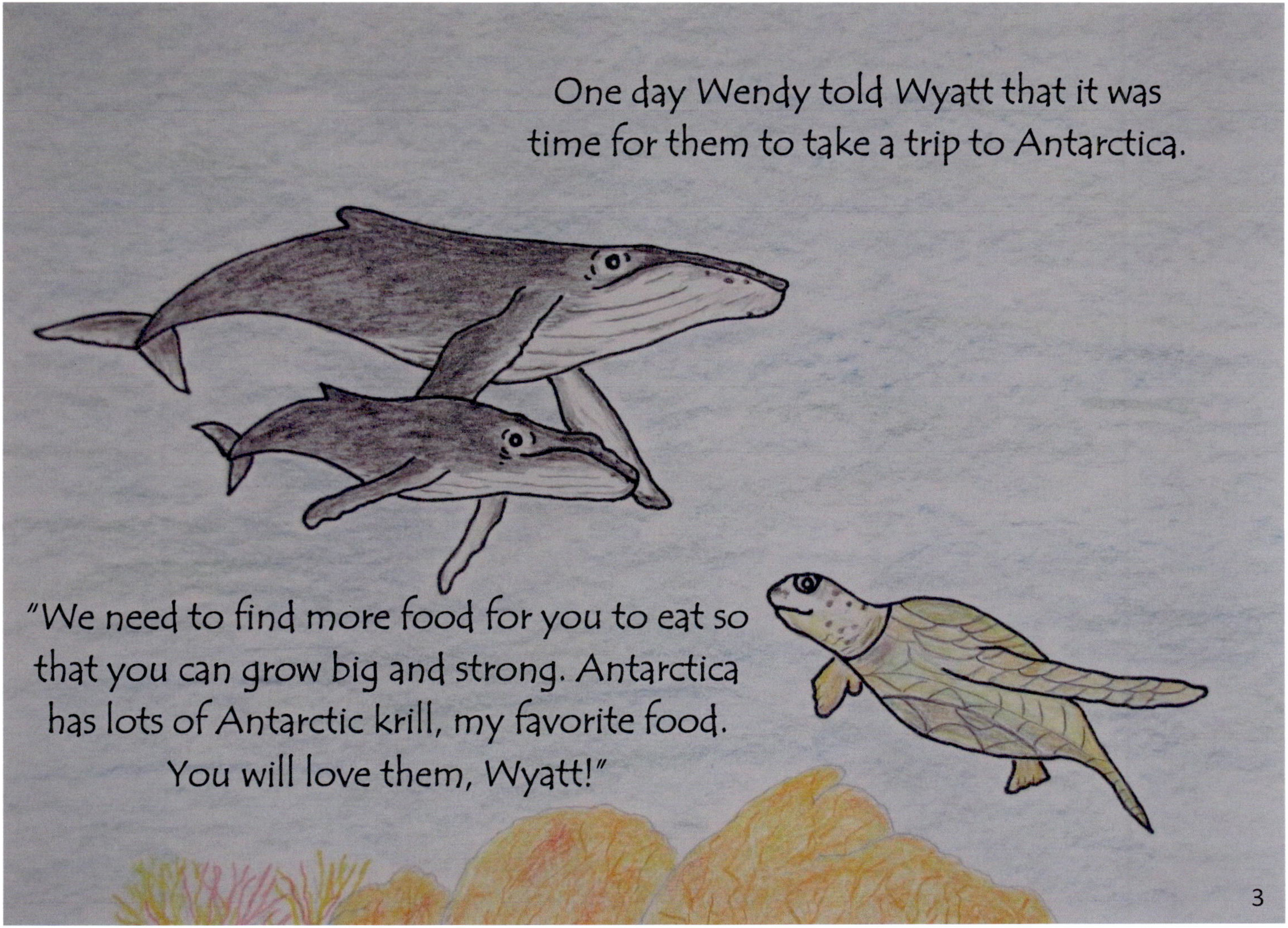

One day Wendy told Wyatt that it was time for them to take a trip to Antarctica.

"We need to find more food for you to eat so that you can grow big and strong. Antarctica has lots of Antarctic krill, my favorite food. You will love them, Wyatt!"

Wyatt had never heard of Antarctica before, but he was getting hungry. There was not much food for him in the tropical waters he called home.

"Antarctica," thought Wyatt. "Sounds nice."

The next day Wyatt and Wendy left the warm waters off of South America and began their trip to Antarctica.

"Wyatt, this trip is our migration" said Wendy. "We will go on this journey every year."

Wyatt was nervous, but excited for this new adventure!

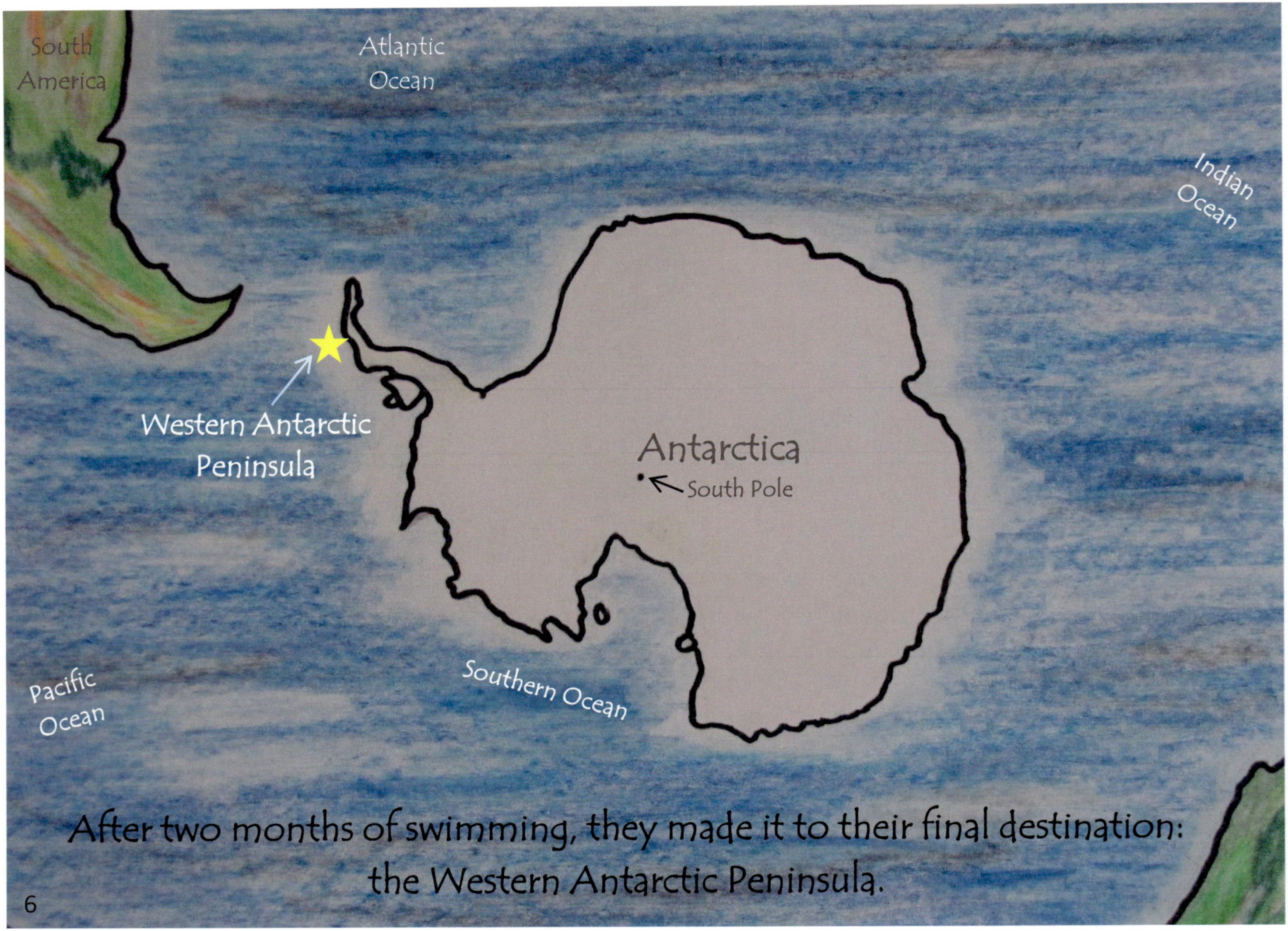

After two months of swimming, they made it to their final destination: the Western Antarctic Peninsula.

The water in Antarctica was very cold, but
the scenery was breathtaking.
Wyatt saw snow covered mountains,
icebergs and glaciers everywhere he looked.

Wyatt also saw lots of other animals that he had never seen before.
He saw

Antarctic fur seals,

crabeater seals,

leopard seals,

minke whales

and gentoo penguins.

He also found lots of Antarctic krill to eat.

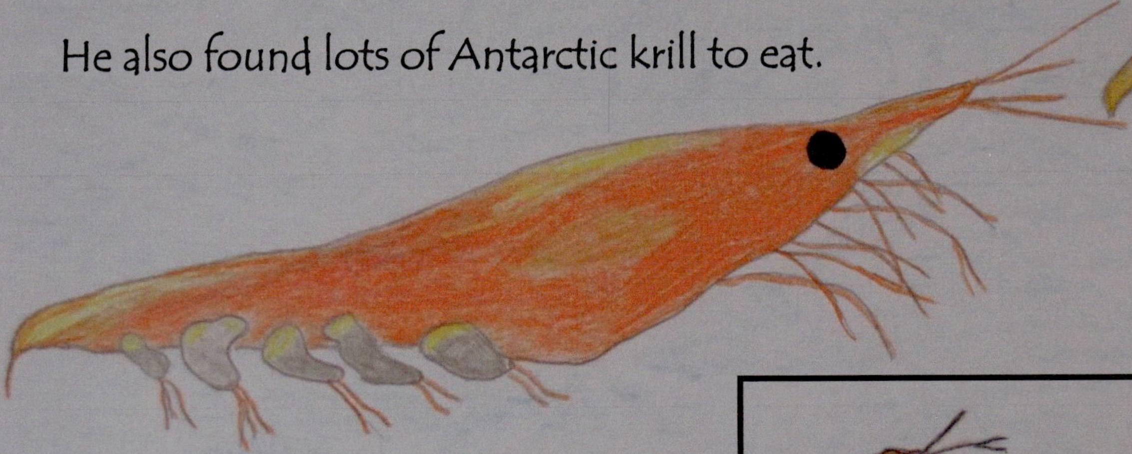

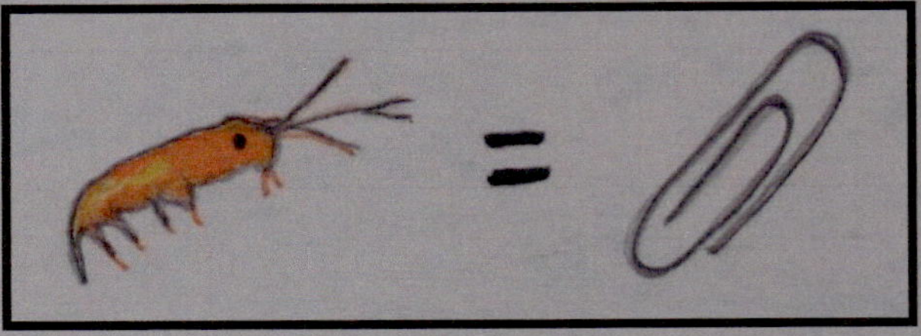

Antarctic krill size = 1 – 2 inches

Because krill are so small, Wyatt was able to eat thousands of them in just a single gulp. He was a very full and happy humpback whale.

One day Wyatt was enjoying some krill when he spotted something unusual in the distance.

"Mom, what is that?" He asked Wendy.

"That's a ship, Wyatt. We see them occasionally here in Antarctica."

Wyatt's mom was right. Wyatt had spotted a scientific research ship. This type of ship carries scientists from all over the world to Antarctica so that they can study things there like ice, rocks and animals.
Little did Wyatt and Wendy know, the scientists on this ship were in Antarctica to study *them*.

"I think I will say hello to it!" said Wyatt as he lifted his fluke out of the water.

A scientist onboard the research ship was looking through her binoculars in hopes of finding some whales to study when Wyatt waved hello.

"We have some whales!" yelled the scientist as she spotted Wyatt fluking.

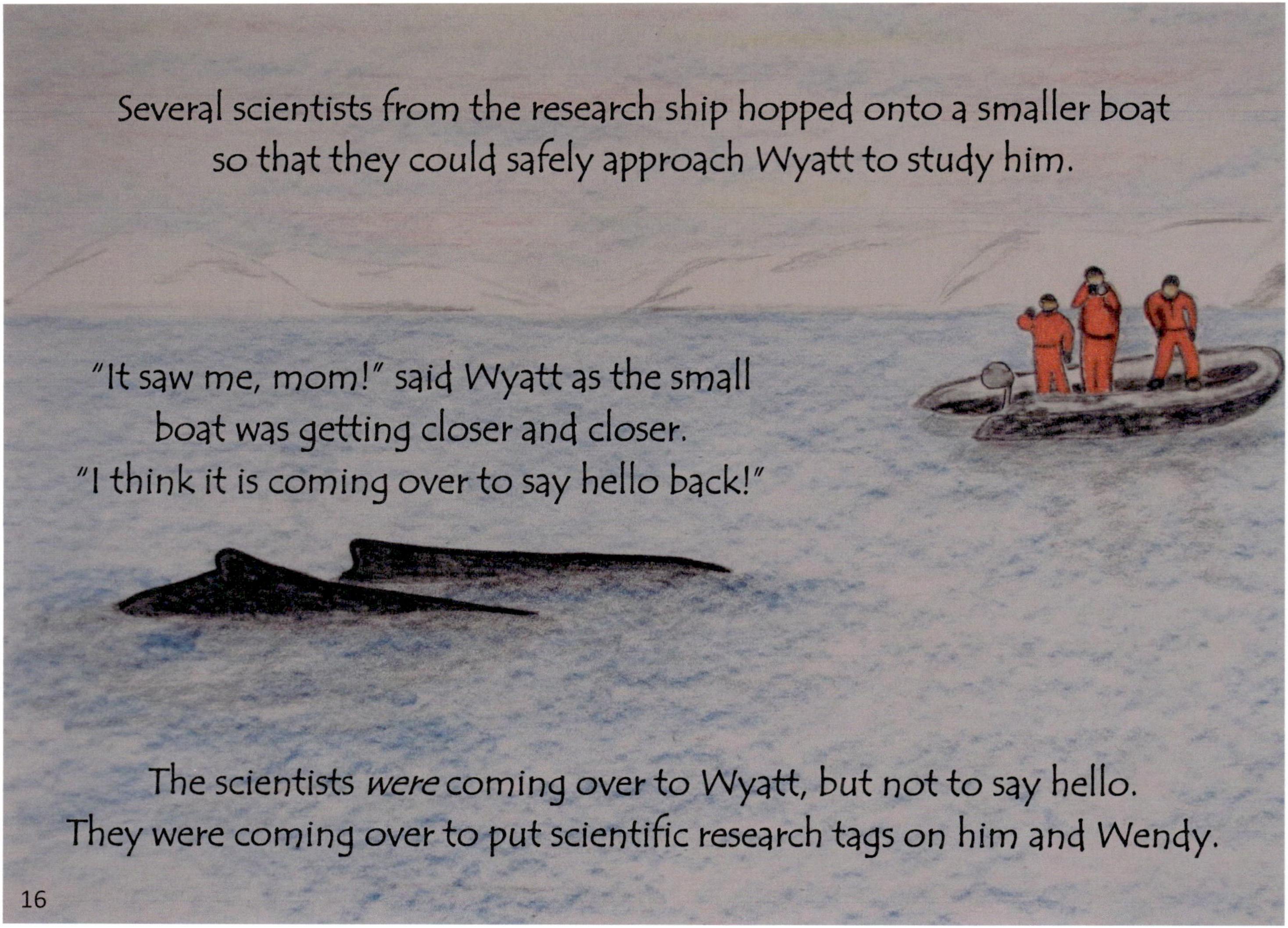

Several scientists from the research ship hopped onto a smaller boat so that they could safely approach Wyatt to study him.

"It saw me, mom!" said Wyatt as the small boat was getting closer and closer. "I think it is coming over to say hello back!"

The scientists *were* coming over to Wyatt, but not to say hello. They were coming over to put scientific research tags on him and Wendy.

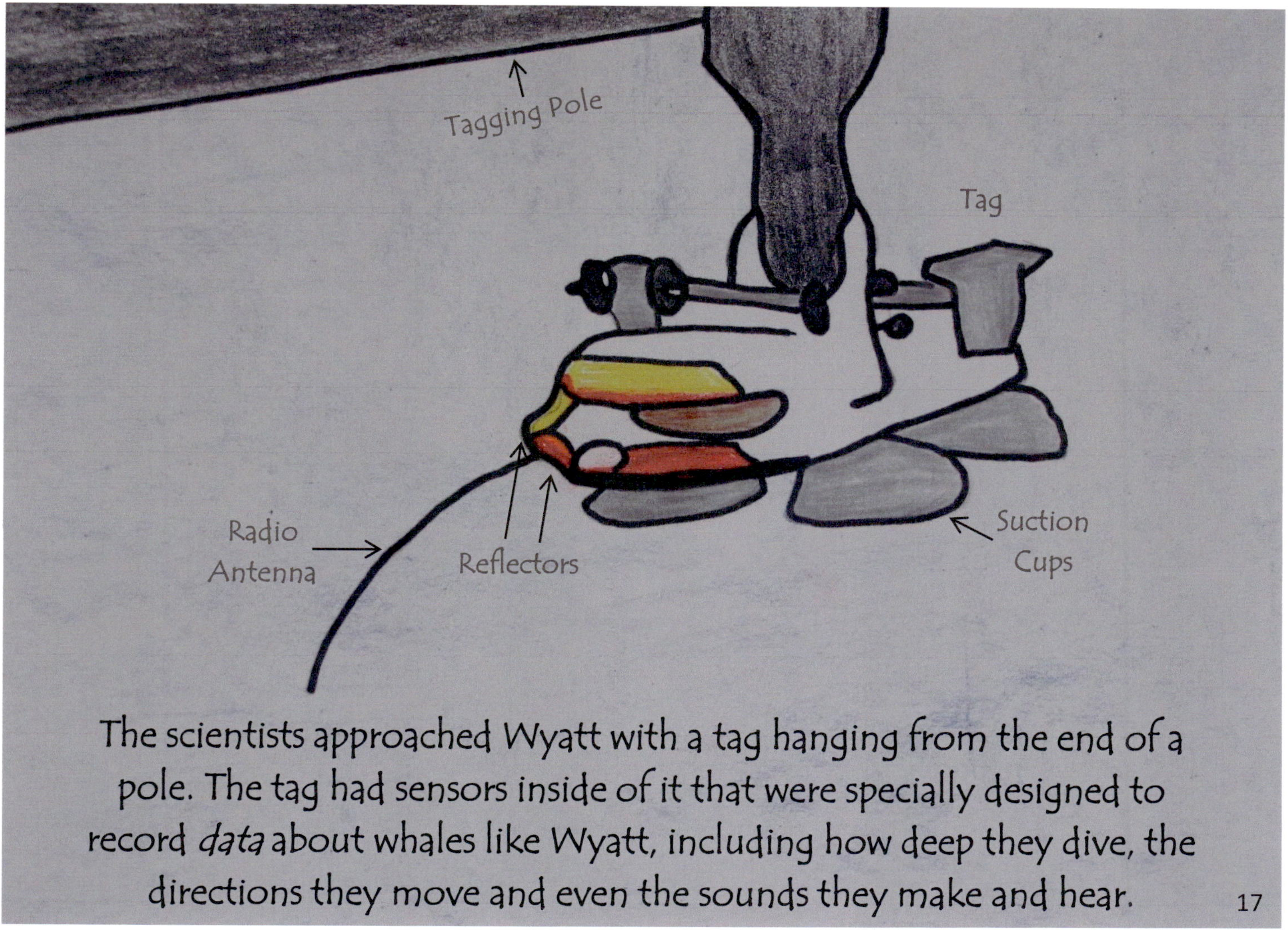

The scientists approached Wyatt with a tag hanging from the end of a pole. The tag had sensors inside of it that were specially designed to record *data* about whales like Wyatt, including how deep they dive, the directions they move and even the sounds they make and hear.

The scientists cautiously approached Wyatt, stuck their long pole out and over him, then placed the tag onto his back in one quick movement.

"Whoa!" Wyatt exclaimed as he felt something tap him. "Mom, I think my new friend just put something on me!"

Wyatt was right! His new friend had put a tag on him.
He had been *tagged* by the scientists.

Before they knew it, another small boat full of scientists was approaching Wendy.

"Mom," yelled Wyatt.
"I think they are going to give you something too!"
Wyatt was right. In an instant, Wendy had also been *tagged* by the scientists.

Wyatt was so excited about the gifts he and his mother received that he lifted his fluke in the air to show it off to all of the penguins and seals floating on icebergs nearby.

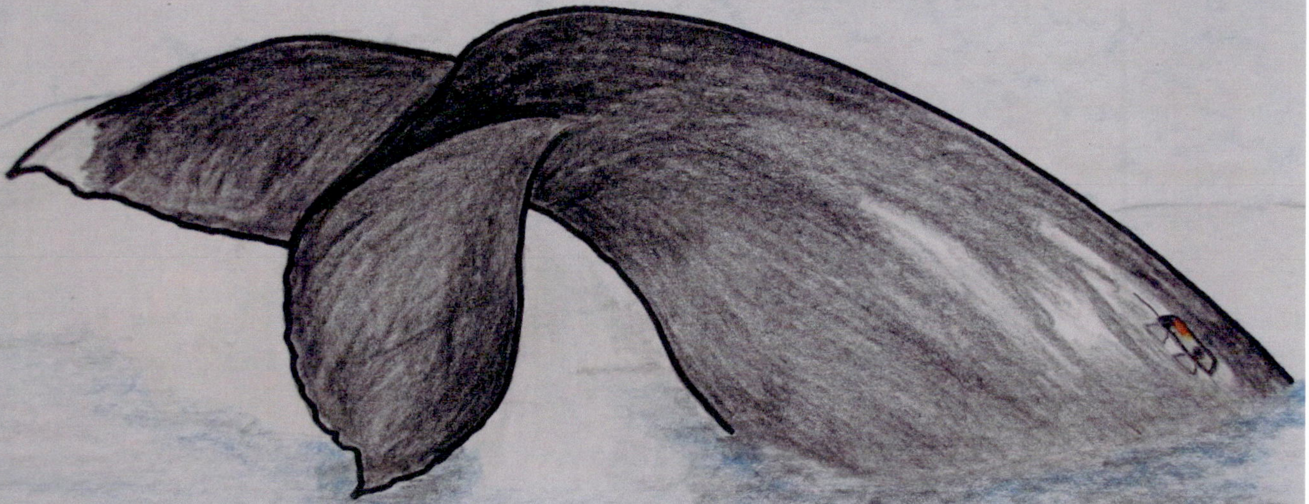

With their new gifts attached, Wyatt and Wendy carried on with their day. They swam around, played games and ate lots of tasty krill. They didn't know it, but these "gifts" were collecting data about them the entire time.

The next morning, Wyatt and Wendy were resting at the surface of the water when the suction cups released and their tags fell off.

"Mom" yelled Wyatt. "I think I lost my gift." He was sad that it was gone.

The scientists were very happy though because now they could download and look at all of the data that the tags collected about Wyatt and Wendy.

Using this data, the scientists were able to create three-dimensional tracks of Wyatt and Wendy underwater.

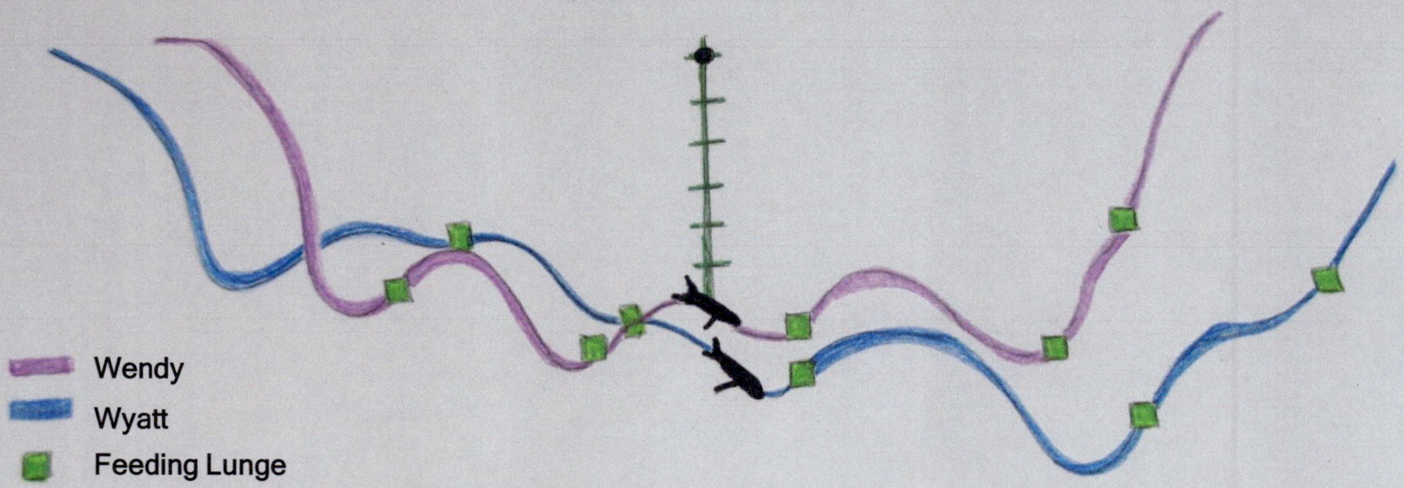

In this picture, Wyatt's behaviors are represented by the blue track while Wendy's behaviors are represented by the purple track. Times that they lunged forward to gulp krill are also shown as green boxes along their tracks. Information like this can help the scientists better understand and protect whales like Wyatt and Wendy.

"Look mom! Our friends look happy!" he said as he saw the scientists jumping for joy. Wyatt was right. This was the first time these scientists had put tags on both a mother and calf humpback whale at the same time. They were very excited about this new data set.

"I'm sad we lost our gifts but I'm so happy that we made new friends, mom!" said Wyatt as he breached high in the air in excitement. He was so happy that they had come to Antarctica.

"Mom, I like Antarctica" said Wyatt a few months later as they started their migration back to South America. "I can't wait to come back next year to make more friends and have even more adventures."

The End

Glossary

- Antarctic krill – Small, swimming crustaceans that live in the Antarctic waters of the Southern Ocean.
- Antenna – An electrical device that can transmit, send and receive radio or satellite signals.
- Binoculars – A tool that allows users to view distant objects.
- Breach – A leap or jump out of the water by a whale causing a great splash in the water.
- Crabeater seal – A medium sized, pale colored seal often found on ice in Antarctica.
- Data – Individual pieces of information collected in a specific environment.
- Download – To transfer data from one system to another (like a computer).
- Fluke - The tail of a dolphin or whale.
- Fur seal - A species of pinnipeds that has external ears, and relatively long, muscular fore flippers.
- Gentoo penguin – Fast swimming birds who are adapted to life in cold, harsh climates.
- Glacier - A body of dense ice formed on land by the accumulation of snow that has melted and frozen over many years.
- Humpback whale - A species of baleen whales that has long pectoral fins, allowing it to be highly maneuverable.
- Iceberg – A large piece of ice that has broken off of a glacier or ice shelf and is floating in the water.
- Leopard seal – The second largest species of seal in Antarctica. It eats other seals, penguins and krill.
- Migration - The physical movement by animals from one area to another.
- Minke whale – The second smallest baleen whale.
- Reflector – A specialized surface used to redirect light towards a given subject so it is visible.
- Scientist – A person who is engaged in and has expert knowledge of a science subject, such as biology or oceanography.
- Suction cup – An object often made of rubber that produces a partial vacuum used to stick an object to a new surface.
- Southern Ocean – Ocean encircling Antarctica, south of 60 degrees of latitude.
- Tag – A tool or device attached to an animal that gathers information about its behaviors.

Did you know?

- Antarctica is the southernmost continent and the fifth largest. It contains the geographical and magnetic South Pole and is surrounded by the Southern Ocean.
- No single country owns Antarctica. Many countries have signed an agreement, called the Antarctic Treaty, to care for and protect it. A significant goal of the treaty is international cooperation in scientific research.
- Marine biology is the scientific study of organisms in the ocean and surrounding waters.
- Oceanography includes the biological, physical, chemical and geological study of the ocean.
- Whales, dolphins and porpoises belong to a group of marine animals known as cetaceans. All cetaceans are mammals that live in the ocean. They breathe oxygen, are warm-blooded and nurse their young.
- Humpback whales migrate up to 12,000 miles each year. They feed in the summer in the cold water of the polar regions. They breed and give birth in warmer waters of the tropical regions during the winter months.
- Markings on the flukes of humpback whales are unique to each individual. This allows scientists to use pictures of humpback whale flukes to identify individuals and track their movements over time.
- Humpback whales feed in a manner known as "lunge feeding." This involves accelerating with a burst of energetic fluking towards and engulfing a huge amount of water and fish or krill. They use baleen, thick hair-like bristles that hang from the top of their mouth, to filter the food from the water they ingested.
- Humpbacks have elastic blubber on the bottom of their mouth that allows them to open their mouth extremely wide and engulf volumes of water and prey that can equal their body size.
- Krill often swim together in large schools, called swarms, which can reach densities of 10,000 – 30,000 individuals per cubic meter (a meter is approximately 3 feet).
- Humpback whale calves stay with their mothers for about one year. They rely on their mother's milk for food at first but learn how to eat krill and other small fish when they get to their foraging grounds.
- On average, humpbacks are 45 feet in length (14 meters) and the females are larger than the males. They can weigh up to 30 tons (about 80,000 pounds).

Made in the USA
Monee, IL
05 November 2026

67382727R00021